THE BOOK OF NNA: II

THE ERA OF DIGITIZED

ENSLAVEMENT

© Kofi Piesie Research Team Same Tree
Different Branch

Kofi Piesie/Mossi Warrior Clan

Copyright 2020 by Kofi Piesie Research Team

Printed in the United States of America

MOSSI WARRIOR CLAN

DEDICATION

In honor of Iya, this book is dedicated to Nna, my arakunrin Chris, my omo Tedros, the great Atsuko Makane Faye, as well as my immediate ebi (Isaiah, Saba, Rahel, and iyawo Zebib) and extended awon ebi like bro Garfield Reid, Ankh West, Kofi Piesie, Ini-Herit Shawn Khalfani, and the Mossi Warrior Clan, the Dagger Squad, Pseudo Killas, Science with Shawn, SaRa Suten Seti, Sheena Lynne, Kecia Jones, Thurston Hargrove, John Pitts and all my supporters on social media as well as all of those who seek truth, wisdom, and understanding and to whom I owe the greatest debt.......... I would also like to thank the medical and emergency teams and the staff from Lawrenceville Emergencies, Gwinnett Medical Center, and Northside Hospital here in the Greater Metropolitan area........ I ask you to remember that.........Tomorrow........is not a given........but yesterday was.......and today.....is being given...........

Those AfRaKaNs who walked the plank

Oh, how I remember that day
When those AfRaKaNs walked the plank
Oh, how I remember Goree
When those AfRaKaNs walked the plank

So many drowned in the Great Ocean
Are those AfRaKaNs who walked the plank
So many kept their devotion
Are those AfRaKaNs who walked the plank

Now many live in the States
as those AfRaKaNs who walked the plank
Must never make the mistake and forget

Those AfRaKaNs who walked the plank

Table of Content

CHAPTER I: The Information Age and a Renewal of AfRAKaN...7

CHAPTER II: How the lack of access to Information Technology (IT) is Destroying Traditional AfRaKa..20

CHAPTER III: Merging IT and Education for the complete Liberation of AfRaKa.........................29

CHAPTER IV: The Importance of AfRaKaN

Investment and the need for Retribution/Reparations.40

CHAPTER V: Why foreign states should transfer Knowledge, Technology and back to AfRaKa................54

CHAPTER VI: The need in AfRAKA for more Technology and Research Centerschnology and back to AfRaKa...63

CHAPTER VII: Using Information Technology to combat the hegemony in international trade relations.71

CHAPTER VIII: Is antagonism the primary catalyst behind the AfRaKaN Information Age.................77

CHAPTER IX: Solar-Powered and Efficient Transportation in the battle against Climate Change...81

CHAPTER X: A Brief Look at AfRaKaN Initiatives to Explore Space..89

CHAPTER XI: CONCLUSION.......................96

CHAPTER ONE

CHAPTER I:

The Information Age and a Renewal of AfRAKaN

At the end of publishing The Book of Nna, I discovered some troubling but liberating news. For most of my existence, I had been told that a person by the name of Jack was my Nna.

After taking a DNA test, I discovered that Jack had told me the truth and that oun was not my Nna. Still, maybe out of sympathy and a sense of inherited AfRaKaN values, Jack continued to maintain a connection with me even though it was a distant one. This was a direct example of inherited cultural values that resonates in most AfRaKaNs in America, whether they are conscious of it or not.

At the same time, while I was unaware of my DNA heritage, I was able to consciously maintain an AfRaKaN perspective in terms of my identity and values thanks to my ebi and the teachings of Dr. Lewis, Central State University, for my educational experiences in general.

In fact, my concept of existence, i.e., being, was defined and driven by the positive and negative aspects from both Jack and my ebi in general.

Something that stood out to me, which I realized much later, was just how significant the development of my skills and knowledge, which included my knowledge of Jack, shaped my consciousness. Jack had experienced AfRaKa and was stationed at a U.S. Navy base known as Kanyo Station in Eritrea. Kanyo Station was a part of the US government's strategy to protect the colonialist control of the flow of water routes coming thru the Suez Canal and the flow of gold and oil, which was being abstracted from what is now referred to as European Imperialism and is a byproduct of the Eurasian perspective and has been imposed on "East" AfRaKa.

My studies as an undergrad motivated me to be more interested in AfRaKaN Etan, which also led me to be particularly interested in Ethiopia, which I had assumed was TaNehesi, Nubia, and Cush. Furthermore, I also developed a strong interest in AfRaKa culture and languages and as well as the etan of AfRaKa in general.

To this extent, my perspective and scholarship were heavily impacted by the knowledge presented to me by the great Hem Neter Dr. Joseph E. Lewis. Under this atsuko's direction, I began to study as much as I could in regard to AfRaKaN history. Atsuko Lewis and other awon atsuko at CSU encouraged my to develop my understanding of AfRaKa. I should mention here that Dr. Lee Ingham and Dr. Jeff Crawford exposed me to philosophy, i.e., The principles of reasoning, logic, and the fundamentals of methodology. This allowed me to become well-prepared and critical in terms of my studies of etan. I concentrated my studies on liberation theology by studying the greats like Dr. Ben and John Henry Clarke and Paulo Freire, and Frantz Fanon. In addition, I should also add other great Hem Neteru such as J.A. Rogers, George G.M. James as well as John G. Jackson.

However, when I met Molefi Asante and was introduced to the Afracentric Idea, I began to focus heavily on methodology, which I see as the natural growth of AfRaKaN thinking and my intellectual development.

In addition to my scholarship, AfRacentricity has always been an essential component of my consciousness. It has made a number of valuable contributions in terms of maturation and the development of my AfRaKan Perspective with the goal of returning to AfRaKa permanently.

My philosophy and methodology studies allowed me to understand the importance of developing a completely independent perspective. I don't claim ownership of it as it is an essential component of the AfRaKan concept of existence.

I decided while in Grad School and teaching at various locations that this approach to scholarship would have to be free from Eurasiancentrism and its poison. I realized that AfRaKa needed to abstract itself from the shackles of Imperialism in order to reach this goal, which would have to include the AfRaKaN contributions to history in addition to an overall perspective that would consist of both knowledge and science.

It was during this time that I also concluded that colonialism had introduced existence to what has been referred to as Existential Imperialism in my earlier books. I must also mention that this led me to a unique understanding and appreciation of just how vital Information Technology (IT) was/is for the restoration of the AfRaKaN perspective on existence.

In this regard, not only does technological development have the potential to lead AfRaKa towards a path of liberation but it also provides a good foundation for preserving traditional AFraKan perspectives regarding existence. What we could then conclude would serve as the primary building block of the AfRaKaN Identity must be fueled by its own traditional perspective.

It is important to remember that we must incorporate modernized values, mores, etc…into our methodology.

I must acknowledge that since opening my mouth after leaving the hospital and therapy, I have experienced a very torturous situation. In fact, after returning to my ebi, I have been treated like "shit" or like a felon by this system which is unfortunate because the interest and goal of educating eniyan about climate change is critical when it comes to preventing an existential level event.

Regardless of the obstacles, I am still committed to promoting awareness of global warming. This motivation has me convinced that IT and science, in general, offers the best possible option for existence to escape the paradigms of intellectual imperialism. These obstacles leave our collective consciousness trapped in shackles of indoctrination that lead to impoverishment, crime, sexual deviance, and an overall decay of modern society.

The objective of this book is to emphasize the value of identity (culture) preservation by adapting technology and a new pedagogy within educational sectors. We will also present the development of my IT skills and how they shaped my concept of TaNeheSi by incorporating Yoruba principles in my method just like I did in The Book of Iya (Vol.1,2,3) as well as Vol.1 of The Book of Nna.

Again, the purpose here has always been the preservation of Traditional AfRaKaN cultural perspective, i.e., a perspective that incorporates indigenous languages, customs, and more of AfRaKa cultures and civilizations. This must include an evaluation of ancient Ta-Merry, Ta-Seti, Ta-Nehesi, DMT, Nubia, and Cushite perspectives. We must understand our responsibility as scholars to be blatantly honest and accurate regarding our views and methodology. The outcome will be the fostering of tools that will assist in the resurgence, restoration and preservation of traditional AfRaKaN culture which by no means is an easy task. It does offer hope and positive outcomes for many, especially among AfRaKa and those who are non-AfRaKaN from an existential perspective. I fear that many have not had the opportunity in the Americas to read my texts and may not understand the perspective presented as it contradicts the mainstream Eurasian paradigm, which, again, is homosexual indeed..........just like the Heka Kasu.

Nevertheless, we must continue to move forward due to the need to address the threat of climate change and its potential to impact our existence as we know it severely. It should also be noted that this is not about any form of religion or any chosen "death-" or "life- "styles but is based on the reality that religion itself is unscientific and has created a vacuum in our consciousness has been usurped by imperialism which results in a lack or dismissal of scientific principles and methodology.

This is a by-product of CaucAsian imperialism and must be addressed if we are serious about the liberation of all things AfRaKaN. Consequently, a critique of the AfRaKaN addiction to religion shows that we have become entirely dependent on and subdued by myth and religious themes that are complete nonsense and contradict the fundamental principles of the scientific methodology, i.e., observation and experimentation as well as the laws of Netcher (nature).

Religion supports belief, whereas science firmly supports observation and experimentation, which sits at the foundation of the scientific methodology and knowledge with conclusions that can be either verified or falsified.

As an IT Specialist, for most of my existence, I was never a faithful follower of religion and obtained a very scientific orientation thru my education and experience, including living on our ebi's farm-life, there, Mathematics, Science, Physics, Math, Logic, and Philosophy, in general, were essential element components of farm life. This led to the formation of rebellion that was incorporated into the AfRaKan Identity and allowed the carving out of an AFRAKaN perspective within the context of the American (HK) experience. I am sure this is the case with other oppressed eniyan in the Americas. Of course, the one who controls the pen also controls the narrative, so it's essential to investigate for yourself.

Instead, far too many of us will not study and may not realize it due to the lack of education. Still, our consciousness has been colonized and subdued by the various forms of pseudo-science and illogical beliefs as opposed to traditional knowledge and cultural systems. The AfRaKaN in America identity has been usurped into the Germanic scheme of "Blackness." You will not find "Black" in AfRaKaN language. So, when the AfRaKaN refers to her/his existence as "black," not only are we colonizing ourselves but opening the door for the colonization of AfRaKa.

There is an attempt to erase the concept of traditional living and existential uniqueness from reality as a whole. One of the dangers of technology, especially the media, is that it can produce a monopoly driven by a monolithic perspective of existence.

If you review traditional AfRaKaN cultural systems, you will see that they are very scientific in nature, and AfRaKa is a pluralistic continue. A key element of AfRaCentific thought/culture is that it allows diversity and pluralism. This is especially true before the invasion by the various clans of the "HK." I am convinced that this conclusion is true for most eniyan and is important and true for most of the other cultures of Geb, including those who may not be AfRaKaN in general. We should allow others to speak for themselves and encourage multi-cultural exchange.

My goal here is not to produce a replica of Eurasian imperialism but to create opportunities for other traditions and cultures to develop and articulate their own perspectives using their own methodologies.

This will allow us to stand firmly and at least be unified in scientific terms. To accomplish this, we will discuss the potential that IT and technology have in general to offer to existence from an AfRaKaN perspective. AfRaKa needs Technology, and while IT provides a lot of promise for AfRAKa, it can only do so if AfRaKaNs take the initiative to escape indoctrination and its colonial origins and pursue knowledge-based initiatives. This also suggests that AfRaKa must not become an imperialist continent but stand firm to defend her traditional pre-colonial identity.

In this regard, the AfRaKaN identity has been attacked, and technology, especially IT, offers the chance to confront the threat head-on. I ask that you reserve judgment regarding the quality or value of the book until you have read all the books in The Book of Iya series, Vol.1 of the

Book of Nna series, as well as other books, including those by my brilliant publishers.

And remember, whether we like it or not, this is shaped and driven by technological interpretation of AfRaKa's development needs with the goal of demonstrating what is needed to extract ourselves from a system of global imperialism that affects much of Geb's population. So, I will demonstrate the promise that technology has in this regard and how we have reached a great moment in Pr Ankh to introduce an AfRaKaN age of renewal in which science and technology are returned to balance.

CHAPTER TWO

CHAPTER II:

How the lack of access to Information Technology (IT) is Destroying Traditional AfRaKa

Recent studies indicate that only 1/3 of Afraka has access to the Internet and other forms of digital technology. Not only does the lack of access impede the technological growth of nations, but it also limits the ability of AfRaKaNs to promote and preserve their own socio-cultural concepts of identity, i.e., their traditional and cultural expressions of existence. An excellent example of this is how the majority of apps are being developed in

African countries with a majority living without electricity access

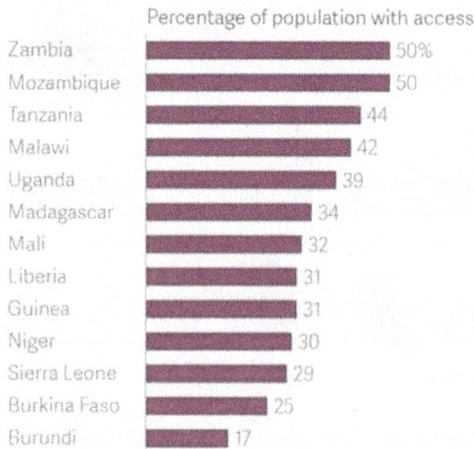

Percentage of population with access

Country	Percentage
Zambia	50%
Mozambique	50
Tanzania	44
Malawi	42
Uganda	39
Madagascar	34
Mali	32
Liberia	31
Guinea	31
Niger	30
Sierra Leone	29
Burkina Faso	25
Burundi	17

ATLAS Data Anabeyonter   Share

English, French, German, Spanish, Chinese, and Korean, and you find many languages tools in such languages but very few in traditional AfRaKaN languages such as Yoruba, Akan, Wolof, Xhosa, Bambara or other Bantu forms of expression.

There are many in the AfRaKaN Geb (excuse me, I meant AfRaKaN Worlid) that concede to the false notion that Hebrew, Arab and other Caucasian languages are parts of traditional AfRaKaN culture and reflect concepts of existence that are AfRaKaN. "Man" and "Wombed Man," as well as "color" consciousness is a perfect example of this. Colonial powers have stripped the AfRaKaN from the power of defining reality in AfRaKaN terms. This includes the current definition of the AfRaKa and how it's been demarcated. ("Worlid") is a perfect example of this failure to view AfRaKA from its own cultural perspective.

What this results in is the complete usurping of the AfRAKaN "being" by an Anglophone or Cauc-Asian linguistic expression of "being" that results in self-imposed imperialism. In addition, we must take into consideration that WEB and Application Development, as well as social media platforms like Youtube, Facebook, and others, are almost completely Eurasian-centered and biased in an imperialist perspective.

When we concede to Caucasian tools of expression, we are also conceding to caucasian hegemony of consciousness, and it's domination and control over much of the epistemology of existence or being as we know it.

Religion and ignorance have left us in a state of vulnerability regarding technological development. For example, how much technology is AfRaKa producing compared to what it consumes, i.e., imports?

Another good example of the hegemony in this technological vacuum is the AfRaKaN dependency on Eurasian concepts of being where instead of utilizing AfRaKaN points of reference referring to ourselves from AfRaKaN perspectives like Yoruba, Akan, Wolofi…etc. AfRaKaNs abroad and at home concede to referring to themselves as "black" or brown. And Eurasian victims of the same tool imposed on states that are dominated by an elite group (caucasian imperialist) will refer to themselves as "white" or "yellow." This denies the victims of imperialism their innate right and power of existence to define themselves according to their own cultural heritage. However, IT allows us to escape this and restore the power of self-definition that traditional societies once had before being colonized.

My studies at Howard and employment at various universities and international development agencies like the UNECA provided me with opportunities to evaluate a number of development issues, such as debt and how those nations that sponsor these loans can hold these clients hostage and use debt as leverage as a means to gain access to AfRaKaN resources. Borrowing is equivalent to a flag of surrender that puts the borrower in a weak position that compromises the recipient state's development agenda.

In this regard, experience at international organizations and initiatives allowed me opportunities to witness the potential imperialist/donor states to manipulate and control colonized eniyan. From an AfRaKaN perspective, one of the more effective ways of reversing the damage caused by imperialism is the supporting AfRaKaN initiatives that are filled with Information Technology and driven by traditional culture. Where unscientific paradigms have left us in a stagnated existence, Technology can be used to offer the digitization of infrastructure as well as to address issues that can be associated with brain drain, depopulation, as well as communication and trade. For example, converting the libraries of Timbuktu and Ancient Kemet into other traditional AfRaKaN languages would allow us to digitize the data and then develop Apps that preserve such content for longevity. Returning to tradition will empower AfRaKa to become a standard and modern for development in a new Information driven society.

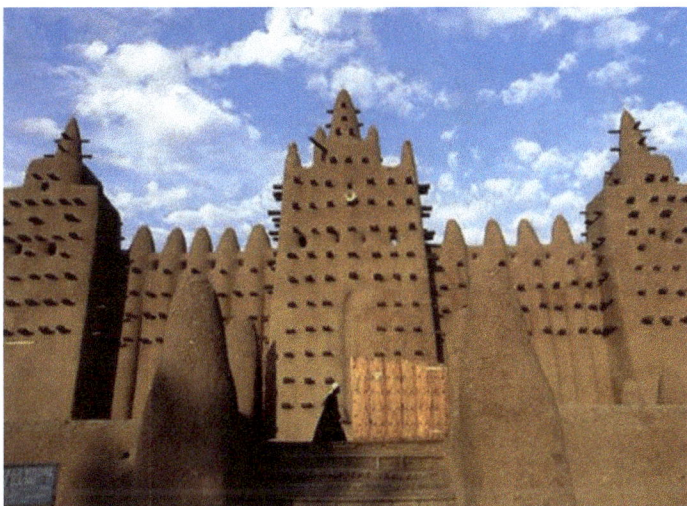

This alone would expand AfRaKaN intellectualism beyond the confines of the shackles and scares imposed on the continent by imperialism and the colonizer's religion. Again, religion is an agent of imperialism.

Based on a review of the chart shown below, it becomes clear that AfRaKa is far behind in terms of technological development.

The World's Most Innovative Countries

2021 ranking of the Global Innovation Index (100 = most innovative)

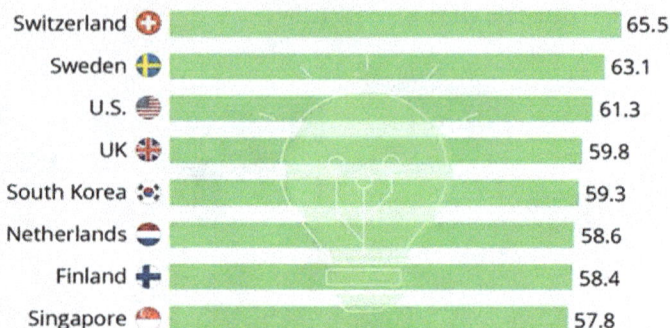

Switzerland	65.5
Sweden	63.1
U.S.	61.3
UK	59.8
South Korea	59.3
Netherlands	58.6
Finland	58.4
Singapore	57.8

Takes into account human capital, institutions, technology and creative output, market and business sophistication, among others
Source: World Intellectual Property Organization

statista

Since the beginning of the colonial experience, the lack of Technology has produced a situation where:

"As of summer 2021, 149 World Bank projects are active in 34 African countries to support the emergence of a vibrant, inclusive and safe digital economy. The projects are diverse and robust and help to narrow the digital divide through electronic identification and banking, mobile money services, family support, training, and mentoring through digital hubs and universities."

This is important because, in the absence of a new information-driven paradigm, access to the information needed to lift AfRaKa and other oppressed eniyan will remain hampered by financial and intellectual poverty. For example

"Too many Africans cannot access the technology they need for school, work, health, or financial services. The pandemic has underscored the vulnerability of the digitally excluded, who have a harder time accessing vital information, including health education and e-commerce.

A 2020 World Bank and GSMA study demonstrated the positive impact of mobile broadband on welfare and poverty reduction in Africa, using data from Nigeria, the largest mobile market on the continent. The percentage of households below the extreme poverty line drops by about 4 percentage points after one year of mobile broadband coverage, and about 7 percentage points after two or more years, in large part due to increased participation in the labor force during that time. Broadband infrastructure is a key driver of jobs and productivity, especially in rural and remote areas."

The important thing to note here is that technology can be an essential aspect of AfRaKa's revitalization as defined by AfRaKaNs. AfRaKa must develop the knowledge of producing their own technology in order to address development needs without being dependent and enslaved by imports.

Currently, what we see exactly is that a few businesses from the Colonizer's camp are using financial systems in favor of a select minority while the majority of eniyan around Geb are subjected to subservient states of existence.

In this regard, it is essential that IT be utilized to re-establish traditional and independent financial systems. This would allow AfRaKa to liberate itself from the global imposition of existential imperialism.

CHAPTER THREE

CHAPTER III:

Merging IT and Education for the complete Liberation of AfRaKa

Much of the struggle regarding the decolonization of AfRaKa has been focused on building wealth. However, while obtaining wealth and financial independence is important and needs to be accomplished as an aspect of AfRaKaN development, the acquisition of wealth can also be problematic as it has the potential to lead many towards greed, nepotism, dictatorship, tyranny, and other forms of exploitation, i.e., moral decay.

This allows the class of select individuals mentioned earlier to become wealthy elitists who can be either foreign or local. For quite some time now, existence for AfRaKa and that of many impoverished nations has been plagued by an international system of exploitation where a minority has benefitted from an allegiance to their imperialist masters who again can be both foreign and local.

Many AfRaKaN "thinkers" like Kwame Nkruma, along with others like George Padmore, Marcus Garvey, and Leopold

Senghor were inspired not only by the philosophical teachings of the abolishment moment in the US but were also influenced by the call for liberation by a global resistance movement. In other words, the same collective energy that produced the call for abolition and decolonization is now demanding reform and repatriation to address existential imperialism, i.e., poverty and the erasing of traditional culture and values.

In response, the imperialists continue to use covert, secretive, and immoral measures to infiltrate various groups who advocate for AfRaKaN liberation both here in the US as a reflection of traditional AfRaKaN values. This can also be applied to the call for justice made by poor Asian and European states.

Patrice La-Mumba, Amilcar Cabral, Steve Biko were all assassinated, while others have been reduced to absolute impoverishment like Marcus Garvey. This results from existential imperialism and the policies of control instituted by the ruling elitists.

It should be noted that the benefactors and primary designers who control this global system are representatives of a financial paradigm that represents and promotes American and Brit-ish interests with the ultimate goal of global domination.

From an AfRaKaN perspective, this provides credence to the perspective that for AfRaKa, education has the potential to serve as a remedy that can and should be used to address issues of dependency and financial and intellectual domination based on a perspective of European intellectualism. Their noted Philosophers like John Locke, Adam Smith, Max Weber, Karl Marx, and Joseph Stalin were all poisoned by their addiction to Eurocentrism and maybe because it is derived from the Caucasian perspective and sits at the foundation of their identity.

This is why a collective expression of resistance has the potential to serve as the greatest opportunity to project and promote a better stance of resistance for oppressed populations in their interests against indoctrination.

However, this will not be an easy task due to the level of penetration of foreign ideas, institutions, and perspectives into the continent. The agents of imperialism continue to be anti-AfRaKan by nature. They are driven by a desire to extract as much resources and wealth from AfRaKaN and Asian populations as possible. Again, it's important for the poor in other groupings to conduct their own research and join other progressive movements by speaking up for themselves without any fear of reprisal.

Furthermore, because these locals are perceived to be either savage or beasts that are incapable of critical thinking and have never made any contribution to collective intellectualism nor contributing anything to civilization as well, it becomes convenient and easy to then justify that such (AfRaKaN) individuals are disposable and can be treated as subservient.

Technology, however, can be a great tool that supports the liberation of AfRaKa from the ignorance and poverty imposed by Asiatic imperialism.

In fact, not fiction, AfRaKan has the potential, just like any other eniyan to produce oun's own technology similar to what our Asian "friends" have been able to accomplish ever since World

War II. We must liberate ourselves from the stigmas of subjugation that suggests that science and technology are in contrast with Netcher.

This is why the ability to demand reparations for both our enslavement and colonialism is very important ,as funding is essential to financial stability and the establishment of IT-based (or centered) educational/learning and application development centers that will advocate and promote our intellectual freedom.

Until this happens, AfRaKa will never be fully liberated so it is important to take steps now that will allow the continent to cleanse out the cronies who control oun's educational systems and who serve as nothing more than agents of a new version of Eurasian colonialism.

Currently, we see a rise in several nations escaping neo-colonial and religious indoctrination in favor of science and technology. It is important to remember that this will not be an easy task but can be accomplished with serious effort and commitment by local populations and their counsels of leadership. This will allow states to adapt strategies that promote their perspectives as well as their traditional cultures.

In this regard, much of the scholarship and intellectualism of AfRaKaN liberation movements, whether on the continent or elsewhere, have been centered around achieving complete liberation, and rightfully so but lack an AfRaKaN-centered methodology.

However, this is not something that can be achieved by education alone because education requires a commitment to funding, of which the modern AfRaKaN state has been deprived of access to financial resources by local and foreign agents.

Furthermore, goods like cocoa and other source/raw products have been exported from the continent at dirt cheap prices while the AfRaKaN imports commercial stage goods like food products, automobiles, computers, diamonds, oil, gas, and other goods that return to AfRaKa with very high tags.

For example, compare the costs of a cocoa bar with that of raw cocoa. As the charts below indicate, the discrepancy is quite disturbing!!

Top Cocoa Producing Countries in the World

Rank	Country	Production (tons)
1	Ivory Coast	1,448,992
2	Ghana	835,466
3	Indonesia	777,500
4	Nigeria	367,000
5	Cameroon	275,000
6	Brazil	256,186
7	Ecuador	128,446
8	Mexico	82,000
9	Peru	71,175
10	Dominican Republic	68,021

FRONTERA
www.frontera.net

Source: UN Food and Agriculture Organization

Africa's Regional Exports Show Ongoing Colonial Legacy

Main export goods of African countries in 2020

Minerals & diamonds
- Gold
- Copper, iron or other ores*
- Diamonds

Energy
- Oil
- Gas

Agriculture, forestry & fishing
- Agricultural products**
- Wood
- Fish

Industry
- Cars
- Boats
- Electrical cables
- Turbines

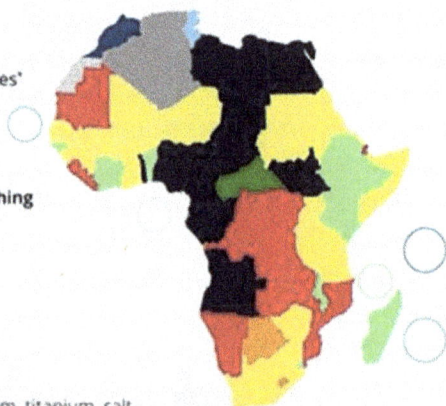

* other minerals: aluminum, titanium, salt
** cocoa, coffee, tea, nuts, spices, cotton, tobacco, essential oils

Source: The Observatory of Economic Complexity (OED)

statista

Merging IT and Education to obtain a Complete Liberation of AfRaKa would liberate AfRaKa from Imperialism and open new avenues for growth and development. AfRaKa should assert its right to trade with other states directly and not rely on foreign parties to negotiate trade deals on behalf of AfRaKa.

These rights sit at the very foundation of essential freedom, and long as other parties have a financial noose around AFraKa's, the continent will remain a victim of imperialism. AfRaKaN pedagogy has been reduced into submission as a result of exploitation. In many regards, this has caused a lack of consciousness in terms of the identity of AfRaKaN peoples worlidwide.

AfRaKa can no longer remain silent to this apparent threat due to global issues such as climate change.

CHAPTER FOUR

CHAPTER IV:

The Importance of AfRaKaN Investment and the need for Retribution/Reparations

While for many, this may indeed be difficult for those entrenched in a Western perspective and seek to "turn the other cheek," AfRaKa has been the victim of significant crimes against the oldest populations around Geb and has been reduced to a state of subservience and oppression which is controlled by the ancient enemies of AfRaKa. This condition was imposed on the collective representation of AfRaKaN existence in general.

The idea presented here is that as "sentient beings," we need to return to our traditional educational values that existed before the invasion during pre-colonial times.

Education can be used as a tool to revert back to perspectives that existed before the periods of imperialism and invasion by the various Caucasian groups, that is to say, the so-called European, his Asiatic cousins, and other cultural and religious groupings such as Muslims, Christians and Hyka Kasu and their break off groupings. The specific focus needs to be placed on the elitist Germanics and their weltanschauung, which inspired rhetoric the "God" trap as an extension of the Cauc-asian perspective.

Restoring traditional AfRaKaN language, culture, social concepts, and science offers a great deal of promise regarding AfRaKaN development. The advance of technology is primarily connected to an escape from religious indoctrination, which should not be taken lightly.

For this to occur, AfRaKa needs funding either in the form of retributive reparations payments or investment by AfRaKaNs and should avoid development loans. However, due to the international system of imperialism, the AfRaKaN continent has been reduced to a condition of impoverishment that is blocked from loans that do not have strings attached. This to is modern enslavement, which keeps AfRaKa trapped in an impoverished shackle. In other words, the continent has inherited a financial system that is a leftover remnant of imperialism and continues to handicap AfRaKaN growth. The AfRaKaN continent has the smell of blood, murder, death, dishonesty, and humiliation all over it, which leads to psychological impediments that are difficult and nearly impossible to dismiss. We must remember that this is a reality that was imposed on the cradle of civilization and not one that has indigenous origins.

From an AfRaKan-centered perspective, it is essential that we formalize a concept of an AfRaKaN reality that is fed by technology which allows IT based Data Centers to collect and store information. These centers should be fully developed and maintained by AfRaKaNs and accessible with boundaries such as censorship. Such an approach will encourage investment that should be used to build high-grade technology centers that can focus on educational and development needs, including internal infrastructure improvements for commerce/e-commerce. A primary justification could be the need to improve the flow and management of AfRaKaN currency. Instead of paper, gold, credit, or digital formats, improving financial infrastructure will allow AfRaKaN states to trade internally and internationally as part fluid and fast past reality.

Such development must be implemented in reasonable terms where they are indeed affordable for AfRaKaN eniyan. Internet access, for example, should be subsidized by the AfRaKAN state, as should education and communication in general. This will produce vibrant environments and lead us to amazing new discoveries and possibilities.

Recent studies have shown that affordability is an impediment to the AfRaKAN adaptation of technology.

"In Malawi, people have been paying around 87% of gross national income per capita for 1GB of mobile data. Rwandans only pay 2%. DW examines the reasons for this huge price discrepancy."

To support IT-based development projects, we must first secure the funding needed to finance such objectives by first arresting and liberating all the wealth stolen from AfRaKa by the agents and cronies of Caucasian imperialism. Many of these agents continue to live in luxury while their eniyan live in absolute impoverishment. Second, the prices of exports such as oil, gas, gold, diamonds, silicon and silicon and others should be raised to international levels and not be paid for in paper money with colonial images or faces on them as this reflects colonial interests, but in Nuba bars (gold).

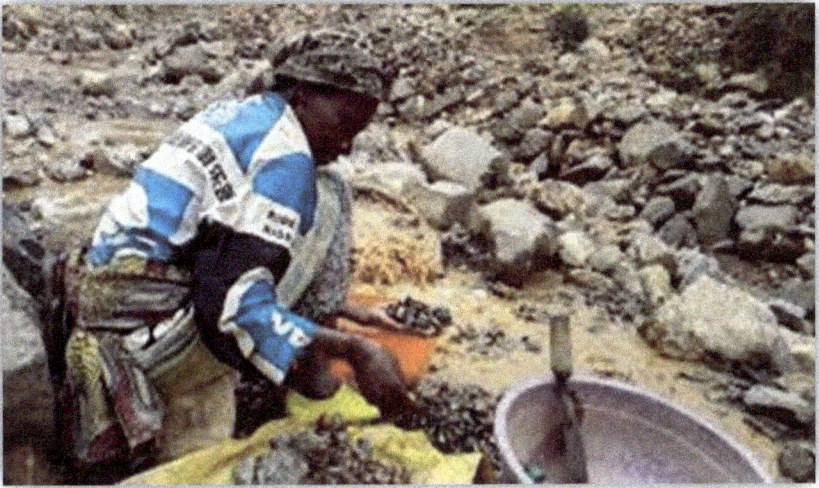

A critical step in this regard is to first facilitate the return of AfRaKaNs to AfRaKa and then create channels for intercontinental trade that will allow equivalent values and goods to be exchanged from country to country on equivalent levels. This approach is derived from the foundation of a traditional AfRaKaN system of communalism and is still more relevant and equitable than the crude materialism characteristic of Western financial systems and has been imposed on AfRaKa and Asia by the agents of imperialism and their cronies.

Due to the perception by many that is volatile, AfRaKaN Data Centers should provide the capacity for Web Hosting and design, along with Virtual services and other hosting solutions such as apps. The importance of this is to liberate AFRAKa financially from control and manipulation like a puppet controlled by strings.

Virtual servers could be used to host online social media platforms similar to Youtube, Vimeo, Facebook, and Roku, which are very popular in the AfRaKaN world. If we can pick up a gun, drive a car, or buy a Computer, doesn't that imply that we should be able to produce our own technology independently? In learning to use a computer, you are, in fact partially what it means to build one.

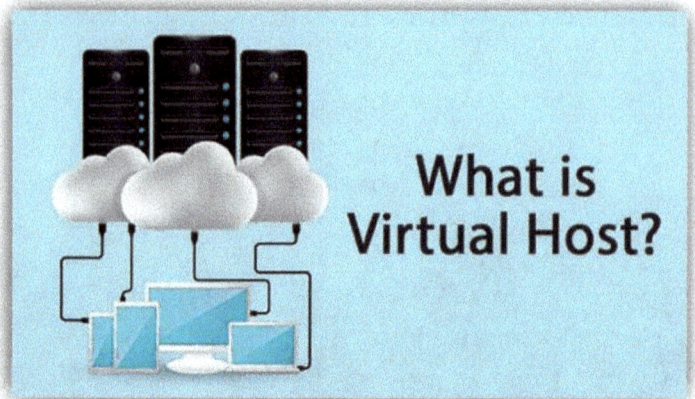

What is Virtual Host?

Offering Cloud space will allow AfRaKaN States to host their internet facilities which would not only provide access to the Internet but also to the engines and applications provided by it, such as Domain registration and content hosting.

"Economic growth in Sub-Saharan Africa (SSA) is estimated at 4 percent in 2021, up from a contraction in economic activity of 2 percent in 2020. However, growth in the region is expected to decelerate in 2022 amid a global environment with multiple (and new) shocks, high volatility, and uncertainty. The economy is set to expand by 3.6 percent in 2022, down from 4 percent in 2021, as it struggles to pick up momentum amid a slowdown in global economic activity, continued supply constraints, outbreaks of new coronavirus variants, high inflation, and rising financial risks due to high and increasingly vulnerable debt levels."

Implementing local communication facilities, including the Internet, will reduce costs and inspire the development of new tools that could be used to enhance exchange regardless of the format. However, as emphasized earlier, the availability of funding is critical.

"Consider this color-based chart which represents Internet tariffs around the continent. Dark green being the cheapest at less than $1, and red being the most expensive at more than $50.

"In Africa, Internet users pay for mobile data more than elsewhere. The disparities in mobile internet tariffs are abysmal, reaching up to more than 1 GB of data."

As indicated below, AfRaKaN connectivity rates are the Highest around Geb. For example, look at Zimbabwe:

"Zimbabwe pays the highest price both in Africa and in the world ($75.20), and countries such as Equatorial Guinea ($65.83) and Saint Helena ($55.47) also pay more than $50. On the contrary, East Africa has the lowest rates, with a gigabyte of mobile data costing ($ 5.93) in Tanzania, ($ 2.91) in Ethiopia, ($ 2.73) in Kenya, (2 $) in Burundi, as well Sudan ($0.68) and the Democratic-Republic of the Congo ($0.88) both pay less than $1 per GB."

"Africa has gotten off on the wrong foot," Amadou Diop, the Senegal-born founder of the digital strategy firm MNS Consulting, says on the topic of digital sovereignty. His company has been hammering out a plan for several months now to address the issue, and Diop paints an alarming picture of Africa's shortcomings on the digital front."

To make Internet access more available and useful in AfRaka, private sector businesses and local investors should be involved and invited to participate parties. This would include local AfRaKaN investors and international agencies like the AfRaKaN Development Bank, UNECA, World Bank, IMF, UNDP, and others to be heavily involved and provide a means to develop nations to reject nepotism and corruption.

Nevertheless, the primary stakeholders and investors should be those reflecting AfRaKaN identities and her perspective(s). As an outcome, this will provide even more of a shield for AfRaKa to protect itself from neo-colonialism and domination by foreign parties. On the technology side, several international agencies have attempted to bring affordable Internet to AfRaKa, which usually has strings attached and has been devastating to local economies as the report below indicates:

"Despite the installation of submarine cables to connect the continent, the price of a gig of mobile internet data remains very high on average."

We should always remember that international funding comes with conditions attached, and AfRaKan states must refuse to be converted into puppets like many states have succumbed to.

DFS funding composition by region

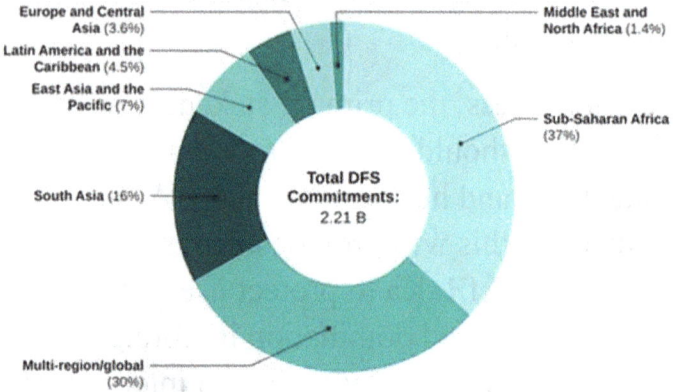

Europe and Central Asia (3.6%)
Latin America and the Caribbean (4.5%)
East Asia and the Pacific (7%)
South Asia (16%)
Multi-region/global (30%)
Middle East and North Africa (1.4%)
Sub-Saharan Africa (37%)

Total DFS Commitments: 2.21 B

n=24 out of 54 funders surveyed with eligible DFS commitments. Data reflects project commitments converted to USD. Other instruments include green bonds, compound bonds, etc.
Source: CGAP Cross-Border Funder Survey 2019

If not properly regulated and controlled, local entities could find themselves facing a huge barrier which has the prostituted puppets leeching off their own eniyan, including the pseudo-religious groups. This is important because there is a goal of erasing traditional AfRaKaN cultural perspectives from the record of Pr Ankh. AfRaKa must collectively confront this issue right on if the AfRaKaN Identity is to be preserved and protected from the space advocates for which AfRaKa has become a resource breadbasket and, in similar terms or by extension, a pawn of Caucasian imperialism in general.

CHAPTER FIVE

CHAPTER V:

Why foreign states should transfer Knowledge, Technology and back to AfRaKa

Reparations would allow AfRaKa to invest heavily in Education, Food production, Medicine, Science, and especially Technology with the goal of being a shining star in the Age of Information and Restoration. And as a result of the expanding use of technology and the appearance of this information age, it is important that AfRaKa asserts its sense of identity within consciousness itself. For example, language demonstrates that imperialistic tongues create more harm than benefit for AfraKan eniyan.

Therefore it is the responsibility of foreign states to reverse their myopic perspective of AfRaKa and approach trade and other forms of exchange on fair and equivalent terms. Already, several non-AfRaKaN countries have begun to invest in developing technology in AfRaKa. It seems that Asian nations like China have been very eager to invest in AfRaKaN infrastructure projects. For example:

"The presence of Chinese multinational enterprises in Africa brings new technology and knowledge. Yet there is a lack of research on the impact of this on development in African countries.

Allowing foreign entities to invest in a country's infrastructure can be very dangerous as it could open AfRaKa up to neo-colonial imperialists and their scheme to dominate the flow of resources on a global scale. In this scenario, only the imperialists benefit significantly, leaving AfRaKaN states behind and in greater debt/poverty.

While not as extreme as Western states, China is a perfect example of this paradigm and should take measures to adjust its trade relationships with AfRaKa on a more equitable basis. As the article shown below, China built itself up out of extreme impoverishment and may have lessons that AfRaKaNs and other subdued eniyan can learn from:

"China's recent industrial development may be more appropriate for Africa's development than Western models, and the adoption of technological knowledge from Chinese firms may provide a more sustainable path to Africa's development because of cultural, institutional, social, and historical synergies."

This comes with challenges from both sides of the trade relationship and stands as an impediment to honest relations or a situation that is genuinely beneficial to peace and stability for all here on Geb. AfRaKa does not need connections with parties that have colonial intentions no matter the reason but must partner with those who are interested and, to some extent, dedicated to furthering what can be labeled "global balance," "harmony" or development.

To this extent, we must thoroughly investigate the intentions of those seeking to invest in AfRaKa, as the continent has been plagued by imperialists who only sought to strip the continent of her resources. For example, like the West, China must be willing to invest in AfRaKa while not being involved in AfRaKaN political and social/cultural affairs. This should not be an issue since China depends on goods and food supplies produced in AfRaKa. However, China just let the West should inherit the perspective of inherent rights and should pay equitable prices for AfRaKa goods.

As the following quotation indicates, China experienced similar historical conditions and managed to institute development policies to raise itself into the age of information.

"However, major barriers may exist to transferring appropriate technology and knowledge. Findings from our study in the construction industry in Ghana suggest an absence of specific technology and knowledge transfer policies and strategies, with human resource development practices, language, and some cultural issues also creating barriers. Bidding practices of Chinese firms investigated also appear to militate against successful technology and knowledge transfers to local partners and staff. Yet there appears to be unrealized potential that has not been addressed by firms. We suggest measures that may be taken to realize this potential and point to implications for policy and future research on the development potential of China in Africa.:"

Like many parts of Asia, the lack of technological and infrastructure development has actually kidnapped AfRaKa and is an issue that needs to be addressed. Technology can lead to numerous possibilities that could be very beneficial for the continent in terms of development. This will spark growth and lead to more equalitarian societies as eniyan moves forward to embrace their traditional heritage. And as a result, a union of culture and technology has the potential to create new roads of possibility that can be free of stigma and dogma and liberate the AfRaKan worlid from the ills of imperialism and ignorance.

In this sense, education, medicine, and health care become even more valuable and vital to AfRAKAN development schemes. Digital apps could be designed and produced at a very lost low cost and become effective means to disseminate information and knowledge as well as scientific literacy.

I have personally benefitted from the miracles of modern medicine and science as my recovery from a life-threatening health issues was facilitated by a combination of both. Currently, my health is being monitored and controlled by medical devices like apps, a defibrillator, a glucose meter, and a smartwatch that I use daily. While recovering, I began to think of how AfRaKaNs are transitioning needlessly due to issues caused by poor health and the inability to afford adequate health care. If anything in this country should be socialized, it's Medicine and Health Care. I also realized that much of what I had gathered to restore my health was from either online resources or apps. Amazon and other digital tools provide a great service that can be used when looking for alternative sources of information.

 The condition of my disability served as the primary foundation and the impetus behind my path of restoration to good health and stability. My IT skills are limited. So as a result, I have not been able or allowed to recover completely. This led me to conclude that in this age of Information, we should use modern technology as a means to improve the condition of eniyan in general, which is a process that can be accomplished by studying traditional languages, medicine, science, and other educational sources. From here, we can develop AfRaKan tools based on AfRaKan needs and not profit. This would include IT tools like database repositories, apps, websites, and others and will incorporate modern scientific principles and methodology.

We need to understand just how critical this is because an

AfRaKaN perspective actually defines our identity and is a critical element in the progression of our psychological health

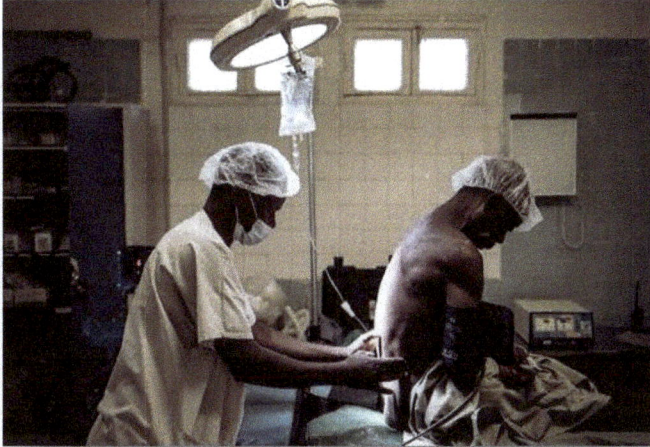

Medicine provides a gateway standard that should be used to measure the level of progress in an Information driven reality.

This would enable us to evaluate the efficiency of governing institutions as well as the quality of leadership within our societies.

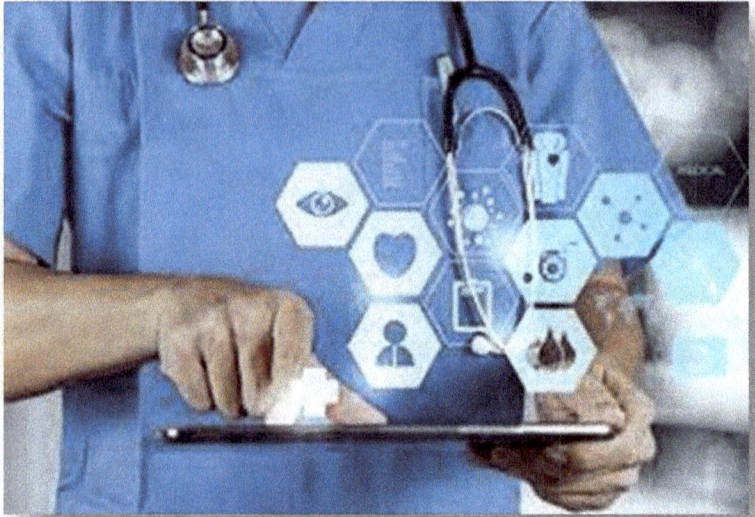

CHAPTER SIX

CHAPTER VI:

The need in AfRAKA for more Technology and Research Centers

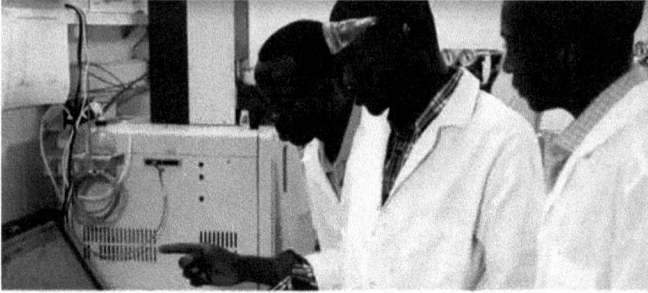

To empower AfRaKaNs to achieve such goals, we should focus our attention and resources on infrastructure development and the construction of Research Centers in this regard. We will evaluate this more in the next chapter by reviewing the achievements made over the past 30 orbits. We will evaluate existing strategies to determine their efficiency and point out opportunities for improvement.

However, let's continue with articulating the justification of Data Centers and the possible benefits that can be derived from them. AfFRaKaNs should own Data Centers that can be placed in local as well as international locations as well as placing storage and communication devices in spaces that reflect traditional AfRaKaN concepts, culture and values.

While many imperialist states use data centers to store information, it collects from populations around the worlid which is often used for espionage and allows imperial (industrial) states to engage in advanced levels of research in the areas of space exploration, weapon production, and medicine. This serves as a wall of protection that now dominates the worlid. For example, as the chart below indicates, these two segments consume the largest percentage of the US budget.

PETER G. PETERSON FOUNDATION — Defense spending accounts for nearly half of total discretionary spending

2020 Discretionary Outlays: $1,627 Billion

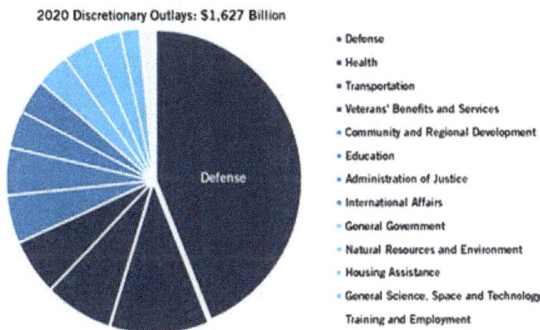

- Defense
- Health
- Transportation
- Veterans' Benefits and Services
- Community and Regional Development
- Education
- Administration of Justice
- International Affairs
- General Government
- Natural Resources and Environment
- Housing Assistance
- General Science, Space and Technology
- Training and Employment

SOURCE: Office of Management and Budget, *Historical Tables, Budget of the United States Government: Fiscal Year 2022*, May 2021.
NOTES: Health includes funding for agencies that provide healthcare services or engage in health research, such as the National Institutes of Health, Centers for Disease Control and Prevention, and Indian Health Service. General government includes central executive and legislative functions as well as the administrative costs of Social Security, Medicare, and income security programs. Energy is included in Transportation and Agriculture is included in Natural Resources and Environment. Veterans' benefits primarily consists of medical and hospital care. In 2020, spending on health programs was boosted by programs to address the pandemic. In 2019, the largest category other than defense was transportation.
© 2021 Peter G. Peterson Foundation   PGPF.ORG

Here's the same budget of the USSR:

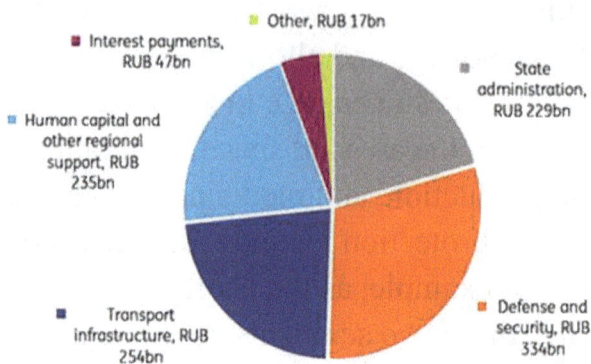

- Other, RUB 17bn
- Interest payments, RUB 47bn
- Human capital and other regional support, RUB 235bn
- State administration, RUB 229bn
- Transport infrastructure, RUB 254bn
- Defense and security, RUB 334bn

And this is a chart that shows how China increased its funding for defense budget by 7.1% in 2022.

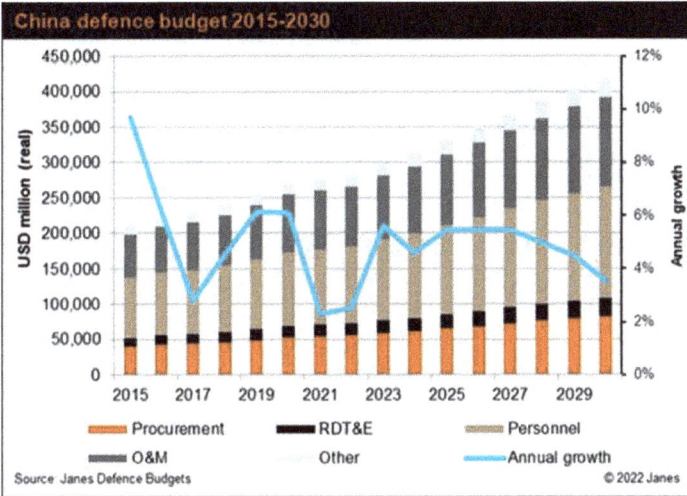

China defence budget 2015-2030

Here's another graph of space countries with the largest budgets:

These countries have the biggest space budgets

Million US Dollars, 2013

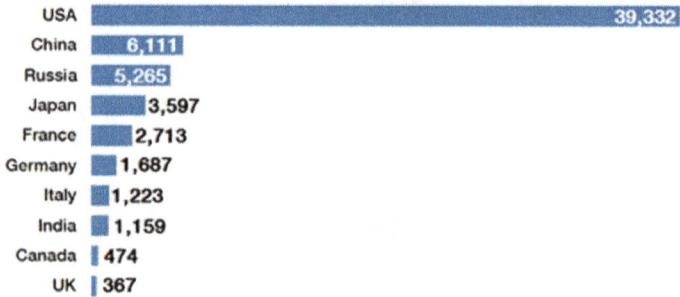

USA	39,332
China	6,111
Russia	5,265
Japan	3,597
France	2,713
Germany	1,687
Italy	1,223
India	1,159
Canada	474
UK	367

Source: OECD

In many regards, it becomes evident that the construction of the modern state is being driven by a goal of supporting the international space program. We have experienced a migration from feudalism to capitalism and from socialism and communism.

These systems accentuate a materialistic expression of existence that is non-Communal and in conflict with Netcher regarding the traditional formation of the state and the infrastructure associated with it. Marxism, for example, is the closest concept Europeans have compared to communalism. However, it continues to remain handicapped by its allegiance to the crude materialism that's present in the Caucasian worlidview.

The underlying component that seems to be the key to them all is using existence as a resource generator that will finance Space exploration. All the financial powers mentioned earlier have thriving Space programs that, in addition to being funded, require a defense. Therefore, existence is paying a price that supports the agenda of a small grouping of Elitists who dominate this reality and have planned an escape.

This explains why Climate change has a huge influence on technological development. But for the poor masses, issues associated with health, food, and the quality of life are more essential questions requiring great attention. At the same time, there is enough wealth and resources to facilitate them both, which suggests that only the greed of the elite few is responsible for the current environment of poverty and hatred that we find almost entirely around Geb. Eliminating this class would allow the creation of a more netcheral world. And in my opinion, technological development is a great tool that can be very useful in terms of development goals.

CHAPTER SEVEN

CHAPTER VII:

Using Information Technology to combat the hegemony in international trade relations

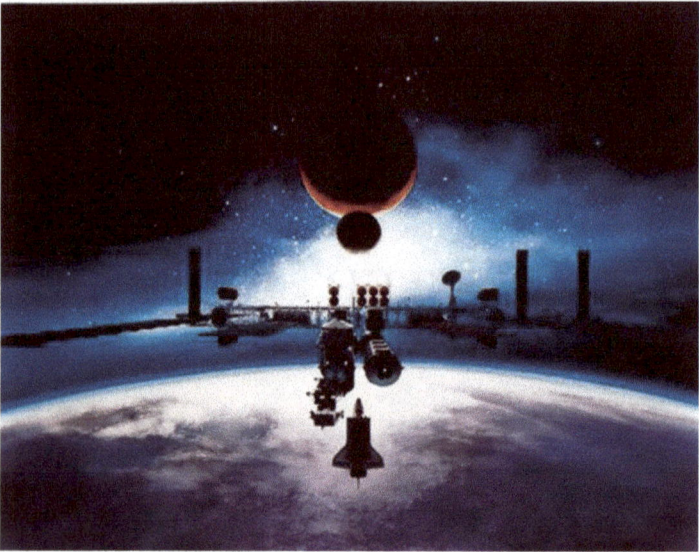

As a Systems Engineer and IT specialist, I was often able to ship PCs we made here in the US back to various countries in AfRaKa. What amazed me the most was that the PCs we would assemble and sell in the States for an average of between $900 and $1500 would sell for $2500 to $3000 in many parts of AfRaKa. Of course, tariffs and shipping costs increased significantly, but the price gap allowed us to understand just how difficult it was to access IT and the lack of local capacity to build them at cheaper rates.

Now some of this was due AfRaKaN crooks and pimps who leeched off of their communities, but when looked at the whole setup, you easily see how it benefitted the materialism of Cauc-asians.

However, as an employee of The Leland Initiative, USAID and UNECA, I was heavily involved in the transfer of technology to AfRaKa. I assisted 52 of 54 countries to connect with the Internet and provided Web development training to many communities. Additionally, I would travel around the continent extensively and set up cyber cafes for business owners, as well as ship computers and assist educational centers in their connectivity needs. These activities were financed by the international organizations I mentioned earlier and what I realized later in my career is that while international donors paid for these financial costs, AfRaKa would pay in terms of her cultural independence and political freedom.

While I was able to accomplish much of the same here in the US, my experience promoting IT always provided me with a sense of intellectual pride and an element of peace and tranquility knowin,g that I could do even more in the land of Iya (AfRaKa).

During that time, many were unaware of computers and the functionality IT had to offer, so the Leland Initiative and the UNECA allowed the team I was affiliated with to become ambassadors of ICTs for development, which was received with a lot of gratitude and appreciation.

Geb doesn't need to be dominated by a few elitists whose errant policies have reduced the quality of existence that keeps Geb in constant conflict. We have been converted into pawns and agents of this self-destructive process as opposed to being agents of modernity,

Unfortunately, many of the oppressed and subjugated worlid continue to be subjects and servants of the international system of Caucasian imperialism, which parades itself as modernization. Furthermore, due to this perspective, many in the developing worlid now find themselves dependent on or addicted to Caucasian sources and rely on this relationship to provide the basic necessities of modern existence. When we need food, medicine, a car, medicine, or education, almost all of these resources are imported from foreign and Caucasian-owned states at very high tariffs.

On the other hand, resources and goods exported from AfRaKa are usually valued as being very cheap. This by itself conditions trade relations and favors the Eurasian seller and not the AfRaKaN buyer. It also favors the Eurasian buyer over the AfRaKan seller on financial terms that can be equivocated to modern enslavement. This implies that an open or free society doesn't exist and is actually nothing more than a slogan used to protect and promote Caucasian hegemony and dominance within financial sectors globally.

In contrast, the culture of submission must be rejected by a powerful renewal of access to information by the oppressed. To protect and promote the Eurocentric perspective, the English language is a covert attempt to converted populations worlidwide into Germanics.

CHAPTER EIGHT

CHAPTER VIII:

Is antagonism the primary catalyst behind the

AfRaKaN Information Age?

The quick answer is no!!! Optimism is an empowering force, so it should be stated for clarity that AfRaKaN advances in technology are motivated primarily by scientific discovery and technological development, which AfRaKa needs to embrace with a serious level of commitment. In fact, the AfRaKaN world is embracing technology even more as it attempts to escape the shackles of imperialism. However, as AfRaKa continues to embrace technology, it must also embrace an Afrakan perspective driven by necessity and not greed. Technology must be accompanied by technical training. This is because before dismissing concepts, we must evaluate cto measure the value and discern whether they provide value for the AfRaKaN reality and can be useful in terms of development.

This also serves as a justification for the creation of a new paradigm, as my awon ebi over at Same Tree Different Branch Publishing have outlined within the concept or perspective of Kimoyo.

On the other hand, we must reject the rhetoric we encounter from all the charlatans and financial pimps from the pseudo camps we encounter as AfRaKaN eniyan. Our development needs require us to focus more on science than indoctrination and religious spookism. Otherwise, the consequence is that we become promoters of imperialists and will promote the wrong perspective for a renewal of AfRaKaN development.

This is in complete opposition to what the elitist HK class has done, which is to use an Asiatic system of mind control to usurp the physical and creative energies of the masses to finance their interests. In contrast, the poor masses pay the bill for their escape route to another planet. A lot of resources are being used to fund the space program with complete disregard for their interest. So just like the Hem Neter Roger Rodney taught in How Europe Underdeveloped

AfRaKa poor states continue to be a basket of resources for Caucasian imperialism which now carried the name of Space Exploration. And instead of telling the scientific truth about the threat of climate change thru mind control, the cauc-asians seem to prefer to use such immoral techniques to finance their own escape.

However, as AfRaKaNs, we must reject all forms of any kind of imperialism so that we can build a new rendition of our traditional AfRaKaN perspective. The challenge here is dealing with the chocolate-chipped oreos who pretend to be AfRaKaN but instead are pawns, agents of the "Father." To this extent, you'll find it much harder to fight against the chocolate chipper than you will regarding a true AfRaKaN and the communities they represent.

In many ways, this shows the importance of an intellectual form of resistance that would stem from eliminating and rejecting the intellectual and mind control techniques deployed by the Elitists around the globe. Priority must be given to discovering traditional AfRaKan language, culture, and existence built on Science and Technology.

CHAPTER NINE

CHAPTER IX:

Solar-Powered and Efficient Transportation in the battle against Climate Change

Climate change is presenting some considerable challenges to this reality. More studies indicate that existence is under serious threat and could be eliminated within a short spectrum of time. To address this threat, a number of measures have either been taken or are in development that has the potential to combat global warming. This includes a transfer to electric vehicles, switching towards eating more health-oriented food, such as vegetarian and Vegan diets, and switching to biodegradable products to reduce the production of waste and garbage.

An issue here is the current cost of food supplies, and climate change has a measurable impact on food supplies. According to a recent report by the International Monetary Fund

"Food supplies and prices are especially vulnerable to climate change in sub-Saharan Africa because of a lack of resilience to climatic" IMF."

Since the goal here has been dominated by profit, fragile farmers do not have the resources or means to protect their crops from the multinational corps (Big Farm); they become victims of GMO farming schemes that contaminate local farming techniques and also eat up the soil and make it more difficult to healthy food supplies.

At the time, the cost of food increased significantly, making it far more difficult for the poor.

"Staple food prices in sub-Saharan Africa surged by an average of 23.9 percent in 2020-22—the most since the 2008 global financial crisis. This is commensurate to an 8.5 percent rise in the cost of a typical food consumption basket (beyond generalized price increases)."

Imagine a world in which climate change, food scarcity, and poverty as insignificant because they is no food to sell or eat. Starvation is so steep that nothing can be done about it but sit, watch, and suffer. International events are leading to the possibility that one of the catalysts for the escalation of the devastating possibilities could be a nuclear confrontation. I personally wouldn't see either, but we do have a group among us who have become so desperate that they could, in fact trigger such an event.

While the possibility of such an event seems rather now given the geo-political state of Geb, we seem to not be able to understand how this impacts markets. In other words, as food availability becomes scarce, the prices of food become more and more expensive, and what the Eurasian systems have seemed to be suggestion that if you don't/can't "work," you will not be paid, and if you are not able to pay then you are not allowed to eat. This means that you primarily become a victim of imperialism and might as well transition, especially if you are a supporter of religion or faith. It seems contradictory, but the whole design of the cauc-asian state is nothing more than a pipe dream. If we take into consideration, we see that global economic and political trends allow us to look deeper into the current crisis and also a further analysis that will enable us to detect just how the Cauc-asian powers are actually using food scarcity and other basic necessities of existence to force communities worldwide into submission. The quote below provides some good insights in this regard:

"The relative strength of a country's currency is another driver as it affects the costs of imported food items. We find that a 1 percent depreciation in real effective exchange rates increases the price of highly imported staples by an average of 0.3 percent."

I am afraid that all of the energy that's being organized to deal with food shortages is actually falling to dead ears and eyes. Awon eniyan doesn't seem to care, nor does it question the rising prices or foot shortages in the market. Instead, attentions are directed at pseudo-intellectualism, a direct descendent of the Germanic (HK)"Black" trap.

AfRaKaN development is crippled by its limitation to non-scientific perspectives that are deeply infused with culture and language, providing the AfRaKaN with tools needed to restore its own view of existence.

https://www.imf.org/en/Blogs/Articles/2022/09/26/africa-food-prices-are-soaring-amid-high-import-reliance

However, climate change is not the only problem farmers face and can be a victim of exaggeration in terms of analysis. Weather trends seem to suggest that it provides a sound justification for investing in knowledge and technology as a remedy.

"Climate change has not been the root cause of the continent's regional famines. Decade after decade, hunger crises have stemmed from a familiar set of circumstances: poor agricultural conditions—including adverse weather, insect infestations, drought, or improper land use—mixed with geopolitical instability. But climate change isn't entirely blameless, either. Like an earthquake that won't stop, climate change breaks foundations that are already weak and also makes it harder to repair and refortify while the ground is continually shaking." IMF

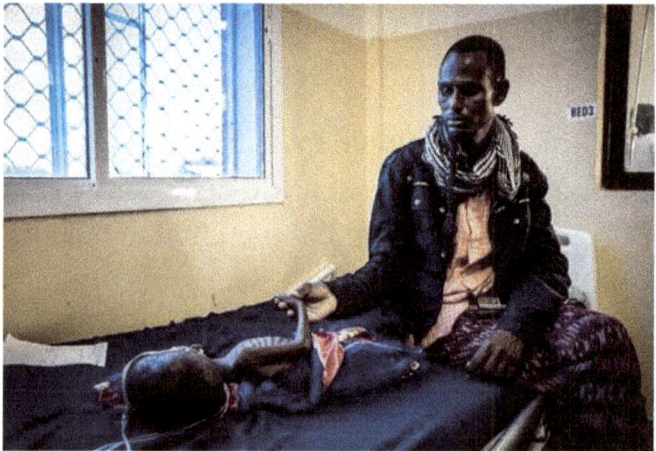

A serious question here is if climate change is an effect, what is the cause? While a portion of this can be blamed on Necher, that doesn't account for the greater portion caused by eniyan.

In my opinion, this is a direct effect of Industrialization that limits the capacity of eniyan to not only deal with climate change but the development needs of modernity. Insect infestation can also present a large threat to a depleted food industry and worsen the pain of famine and other crises. In the ancient Tanehesi/Demet/Nubian/Cush portion of AfRaKa:

"The current swarms of locusts are the most serious outbreak in decades, affecting Somalia, Ethiopia, Kenya, Eritrea, and most recently, Tanzania and Uganda. The locusts could fly into South Sudan, a country whose population is already struggling with high levels of hunger. The swarms are huge, in North-Eastern Kenya, one swarm was measured to be 37 miles long and 25 miles wide."

What this seems to suggest is that insect infections can cause just as much lethal damage as climate change or drought. It must be noted that AfRaKaN is not a safe haven against global crises but serves as the safest location to shield AfRaKANs from too much exposure to existential imperialism.

Yet, because so many of us are caught in a spelled cast reality, we can't see.......... actually, we refuse to see what's going on and actually contribute to the problem. Instead of existing in a perpetual state of ignorance and submission, we must take a unified stance to demand our awon ebi and our dignity. In addition to this, traditional AfRaKaN was highly based on farming, but like other segments of the economy, much of this capacity has been controlled or wiped out by Eurasians and mega-farms. AfRaKa has become the breadbasket of both the Caucasian and the Asiatic.

"Around 45% of food supply in South Africa is wasted, impacting the economy, water security and contributing to climate change, according to research."

CHAPTER TEN

CHAPTER X :

A Brief Look at AfRaKaN Initiatives to Explore Space

The underlining that is driving the AfRaKaN interests in space exploration can actually be placed in two camps. One is based on a sense of understanding the discoveries of modern science. The other is understanding the changing times and the need to rid the planet of the so-called "White" elicits.

I think a third option is also available, and that's where you don't choose either side and decide the method is questionable the objective of colonizing another planet is the best option if existence will continue to be present on this planet.

Therefore, from this perspective, space exploration has merit, but since most of, while not have the means to go or no intention of leaving Geb like me, I think the Cauc-Asian should subsidize the construction of his grand escape route with his own resources his or when he needs them from others.

I've been doing a lot of research since I purchased a new telescope and have learned that AfRaKaNs are active participants in this rapidly advancing branch of science; for example, many AfRaKaN countries have launched satellites into space and have established space programs. According to The Economist:

"The small cube—Nigeria's first satellite and only the second launched by a sub-Saharan African country—did not just watch a storm; it provoked one, too. British politicians and a taxpayers' pressure group called for a halt in development aid, saying Nigeria did not need help if it could afford a space program. Still, the sums being spent on space by African countries back then were tiny. South Africa's SUNSAT, the region's first satellite, was built by students at Stellenbosch University and hitched a free ride on a NASA rocket. Nigeria's spacecraft cost just $13m."

As the article below demonstrates, how Nigeria's persistence has allowed the state to push forward with the creation of a space agency and demonstrates that the Caucasian states will not have a monopoly control over space exploration and that it is better to engage in multiple perspectives when it comes to the potential to discover another location where existence might and could be possible.

Due to the potential of climate change to produce an extinction event, it is good that AfRaKaN states are getting involved in this emerging area of science. Because to avoid what could be an extinction-level event, the international community has to step up to the table and demonstrate a solid commitment to space exploration and AfRaKaN participation on multiple levels. AfRaKaN material resources are not the only area where AfRaKan contrition would be beneficial. AfRaKaN perspectives offer a different approach to scientific exploration as it incorporates concepts of balance and harmony which will be extremely valuable during long space voyages.

The best way that AfRaKa can add value to space exploration is to provide training and assistance that will allow space pioneers to overcome the limitations of their addiction to crude materialism. Dealing with the threat of climate change as a unified team would help increase the possibility of success in space missions.

Jun 17th 2021

We've seen several cases where we failed to
take a unified stance voluntarily to promote and
project a collective sense of commonality. The
American experience has provided us with a
number of perfect examples, like the response
to Hurricane Katrina. The support from
AfRaKa is not well known by many, and we
choose instead to tear AfRaKa down. Yet, we
do have a record of AfRaKaN involvement in
supporting relief efforts. This demonstrates the
ability of AfRaKaN states to provide strong and
adequate levels of technical support in terms of
scientific research. Consider the quote below
for example:

"IN THE HOURS after Hurricane Katrina slammed into America in 2005, destroying large parts of New Orleans, the people coordinating the disaster response urgently needed satellite pictures to show them what they were facing. The first images to come in were not from the constellations launched by NASA or the space agencies of other rich countries. They were beamed to Earth by a small Nigerian spacecraft launched from Russia just two years earlier." (5)

With climate change becoming a global threat and with mostly Cauc-Asians being affected, I think many eniyan are skeptical and hesitant. So the question becomes why should non-cauc-Asians actually care, and what keeps AfRaKaN and others from reciprocating with a similar expression of anger and hostility that many of us have experienced throughout modern history?

We have limited options to being usurped into a wide-ranging initiative to conquer the Geb with mental indoctrination of eniyan. In this regard, moving forward with AfRaKaN expressions of technological advancement is the best option, given the existential condition of the continent where resources have been depleted, and poverty rates are high.

The threat and impact of global warming suggest that investment in space exploration is a necessity and not an option. This may explain why Nigeria has invested heavily in space exploration as well while dealing with the potential financial instability.

I suspect that the level of investment has increased due to recent weather trends and rising costs. Yet AfRaKaN space programs are actually popping us in many locations around the continent.

"In the past few years, however, the continent has dashed into space. Mauritius, the most recent orbital enthusiast, put up a satellite on June 3rd. At least 20 African countries now have space programs. These include heavyweights like Egypt, Algeria, and Nigeria and smaller countries like Ghana. In 2019 another five African countries launched satellites, bringing Africa's total in orbit that year to 41." (6)

We will have to put our historical grievances aside momentarily to prevent the possible extinction of existence here on Geb.

CHAPTER ELEVEN

CHAPTER XI: CONCLUSION

AfRaKa should become more adamant about the Retribution owed by Colonial Powers and the protection of all her resources from exploitation. Instead, the continent is in a ripe position to educate its youth so they can become more involved and active in science and technological development.

However, to accomplish this objective, the imperialist states must also return AfRaKa's stolen eniyan and her Nuba Bars.

Simultaneously, you AfRaKaNs must stand up and challenge the cultural perspectives imposed on AfRaKaN and restore the traditional views of those who walked before us.

The continent has much to look forward to as the influence of imperialism continues to decay, and this opens new paths for development. As a cornerstone, technology should be fully embraced. This masked a focus on ICT and App development as valuable tools for restoring the traditional AfRaKaN perspective regarding existence. In closing, we should always remember that.........Tomorrow...........is not a given..........but yesterday was.........and today.........is still..........in the process of being given!!!

Bibliography

Quotations

(1).https://www.imf.org/en/Blogs/Articles/2022/09/14/how-africa-can-escape-chronic-food-insecurity-amid-climate-change

(2).https://www.cips.org/supply-management/news/2021/august/almost-half-of-south-africa-food-supply-is-wasted-says-study/

(3).https://www.imf.org/en/Blogs/Articles/2022/09/26/africa-food-prices-are-soaring-amid-high-import-reliance

(4).https://www.nytimes.com/2005/08/30/us/hurricane-katrina-slams-into-gulf-coast-dozens-are-dead.html

(5). The Economist:
https://www.economist.com/middle-east-and-

africa/2021/06/17/africa-is-blasting-its-way-into-the-space-race

(6). The Economist: https://www.economist.com/middle-east-and-africa/2021/06/17/africa-is-blasting-its-way-into-the-space-race

 (7). https://time.com/5784323/un-locust-east-africa/

(8). https://time.com/6220057/climate-change-africa-food-crisis/

Books

Afrika, Llaila Melanin What Makes Black
People

Black! New York: Seaburn Publishing Group,
2009

AfRaKan Academy of Sciences. Workshop on
Science and Technology Communication
Networks in AfRaKa. Nairobi: AfRaKan
Academy of Science, 1993

Balaam, David N., and Michael Veseth, eds.
Introduction to International Political Econo
New Jersey: Prentice Hall, 1996

Baradat, Leon P. Political Ideologies. New
Jersey: Prentice Hall, 1994

Billet, Bret L. Investment behavior of
Multinational Corporations in Developing

Areas. New Brunswick: Transaction Publishers, 1991

Clough, Michael. Free at Last?. New York: Council on Foreign Relations Press, 1992

Dalley, Stephanie. Myths from Mesopotamia. New York: Oxford Univeristy Press, 1992

Drew, Eileen P., and F. Gordon Foster, eds. Information Technology in Selected Countries. Tokyo: United Nations University, 1994

Dubois, W.E.B. The World and AfRaKa. New York: International Publishers, 1965

Fieldhouse, D.K.. Black AfRaKa: 1940 1980. Boston: Unwin Hyman, 1986

Haggard, Stephan, and Robert R. Kaufman, eds. The Politics of Economic Adjustment. Princeton: Princeton University Press, 1992

Harbeson, John W., and Donald Rothchild, eds. Afiica in World Politics. Boulder: Westview Press, 1991

Leedy, Paul D. Practical Research: Planning and Design . 5th ed. New York: Macmillan Publishing Company, 1993

Men-Ib Iry-Maat, Wudjau. A Beginner's Introduction to Medew Netcher 2nd ed. USA Wudjau Men-Ib Iry-Maat, 2016

Moran, Theodore H. Multinational Corporations. Massachusetts: Lexington Books, 1985

National Research Council, Office of International Affairs, Bridge Builders. Washington: National Academy Press, 1996

Obenga, Theophile. African Philosophy. USA: Brawtley Press, 2015

Piesie, Kofi. Beautiful Lessons About Kimoyo USA: Same Tree Different Branch Publishing, 2021

Piesie, Kofi. Spear Masters, USA: Kofi. Piesie Research Team, 2021

Rodney,Walter. How Europe Underdeveloped AfRaKa. Washington: Howard University Press, 1974

Reid, Garfield. Misconceptions & Misinformation by the Black Hebrew Israelites Vol 1. USA: Garfield Reid, 2021

Rosenbloom, Richard. Technology and Information Transfer. Boston: Harvard University Press, 1970

Sandbrook, Richard. The Politics of AfRaKa's Economic Recovery. New York: Cambridge University Press, 1993

Segal, Ronald. Islam's Black Slaves. New York: Farrar, Straus and Giroux, 2002

Shibre, Zewdie, and Abdulhamid Bedri, eds. Regional Development Problems in AfRaKa . Addis

Ababa: Institute of Development Research, 1993

Slater, Robert O., Barry M. Schutz, and Steven R. Dorr, eds., Global Transformation and the Third World. Boulder: Lynne Rienner Publishers, 1992

Steindorff, George and Seele, Keith C. When Egypt Ruled the East. Chicago: The University of Chicago Press, 1942

Turabian, Kate. A Manual for Writers. 5th ed. Chicago: University of Chicago, 1987

Weiss, Thomas G., and Merl A. Kessler, eds. Third World Security in the Post Cold War Era

Boulder: Lynne Reinner Publishers, 1991

Weston, Alan F. Information Technology in a
Democra . Cambridge: Harvard University
Press, 1971

Articles, Papers and Public Documents

da Costa, Peter. "AfRaKa Communication:
Internet   A Statist Model", Addis Ababa:
International Press Service, 10 September 1996,

National Telecommunications and Information
Administration, "U.S. Goals and Objectives for
the Information Society and Development
Conference", prepared remarks of Vice
President Al Gore, delivered via satellite to the
Information Society and Development
Conference in Midrand, South AfRaKa(May
13, 1996)

Semret, Nemo. "Unleashing AfRaKa's
Potential: The Technological Reasons for Open
and Competitive Cybercommunications", a

paper delivered at The Second Annual Meeting of the AfRaKa Scientific Society, Washington, June 22, 1996

Burka, Lauren P. "A Hypertext History of Multi-User Dimensions." MUD History. 1993. http://www.utopia.com/talent/lpb/muddex/essay (2 Aug. 1996).

Fine Arts." Dictionary of Cultural Literacy. 2nd ed. Ed. E. D. Hirsch, Jr., Joseph F. Kett, and James Trefil. Boston: Houghton Mifflin. 1993. INSO Corp. America Online. Reference Desk/Dictionaries/Dictionary of Cultural Literacy (20 May 1996)

AfRaKan Academy of Sciences. Workshop on Science and Technology Communication Networks in AfRaKa. Nairobi: AfRaKan Academy of Science, 1993

Balaam, David N., and Michael Veseth, eds. Introduction to International Political Econo New Jersey: Prentice Hall, 1996

Baradat, Leon P. Political Ideologies. New Jersey: Prentice Hall, 1994

Billet, Bret L. Investment behavior of Multinational Corporations in Developing Areas. New Brunswick: Transaction Publishers, 1991

Clough, Michael. Free at Last?. New York: Council on Foreign Relations Press, 1992

Dalley, Stephanie. Myths from Mesopotamia. New York: Oxford Univeristy Press, 1992

Drew, Eileen P., and F. Gordon Foster, eds. Information Technology in Selected Countries. Tokyo: United Nations University, 1994

Dubois, W.E.B. The World and AfRaKa. New York: International Publishers, 1965

Fieldhouse, D.K.. Black AfRaKa: 1940 1980. Boston: Unwin Hyman, 1986

Haggard, Stephan, and Robert R. Kaufman, eds. The Politics of Economic Adjustment. Princeton: Princeton University Press, 1992

Harbeson, John W., and Donald Rothchild, eds. Afiica in World Politics. Boulder: Westview Press, 1991

Leedy, Paul D. Practical Research: Planning and Design . 5th ed. New York: Macmillan Publishing Company, 1993

Men-Ib Iry-Maat, Wudjau. A Beginner's Introduction to Medew Netcher 2nd ed. USA Wudjau Men-Ib Iry-Maat, 2016

Moran, Theodore H. Multinational Corporations. Massachusetts: Lexington Books, 1985

National Research Council, Office of International Affairs, Bridge Builders. Washington: National Academy Press, 1996

Obenga, Theophile. African Philosophy. USA: Brawtley Press, 2015

Piesie, Kofi. Beautiful Lessons About Kimoyo USA: Same Tree Different Branch Publishing, 2021

Piesie, Kofi. Spear Masters, USA: Kofi. Piesie Research Team, 2021

Rodney, Walter. How Europe Underdeveloped AfRaKa. Washington: Howard University Press, 1974

Reid, Garfield. Misconceptions & Misinformation by the Black Hebrew Israelites Vol 1. USA: Garfield Reid, 2021

Rosenbloom, Richard. Technology and
Information Transfer. Boston: Harvard
University Press, 1970

Sandbrook, Richard. The Politics of AfRaKa's
Economic Recovery. New York: Cambridge
University Press, 1993

Segal, Ronald. Islam's Black Slaves. New
York: Farrar, Straus and Giroux, 2002

Shibre, Zewdie, and Abdulhamid Bedri, eds.
Regional Development Problems in AfRaKa .
Addis

Ababa: Institute of Development Research,
1993

Slater, Robert O., Barry M. Schutz, and Steven
R. Dorr, eds., Global Transformation and the
Third World. Boulder: Lynne Rienner
Publishers, 1992

Steindorff, George and Seele, Keith C. When Egypt Ruled the East. Chicago: The University of Chicago Press, 1942

Turabian, Kate. A Manual for Writers. 5th ed. Chicago: University of Chicago, 1987

Weiss, Thomas G., and Merl A. Kessler, eds. Third World Security in the Post Cold War Era

Boulder: Lynne Reinner Publishers, 1991

Weston, Alan F. Information Technology in a Democra . Cambridge: Harvard University Press, 1971

Articles, Papers and Public Documents da Costa, Peter. "AfRaKa Communication: Internet   A Statist Model", Addis Ababa:International Press Service, 10 September 1996,

National Telecommunications and Information Administration, "U.S. Goals and Objectives for the

Information Society and Development
Conference", prepared remarks of Vice
President Al Gore, delivered via satellite to the
Information Society and Development
Conference in Midrand, South AfRaKa(May
13, 1996)

Semret, Nemo. "Unleashing AfRaKa's
Potential: The Technological Reasons for Open
and

Competitive Cybercommunications", a paper
delivered at The Second Annual Meeting of the
AfRaKa Scientific Society, Washington, June
22, 1996

Burka, Lauren P. "A Hypertext History of
Multi-User Dimensions." MUD History. 1993.
http://www.utopia.com/talent/lpb/muddex/essay
(2 Aug. 1996).

Fine Arts." Dictionary of Cultural Literacy. 2nd
ed. Ed. E. D. Hirsch, Jr., Joseph F. Kett, and
James Trefil.

Boston: Houghton Mifflin. 1993. INSO Corp.
America Online. Reference
Desk/Dictionaries/Dictionary of Cultural
Literacy (20 May 1996

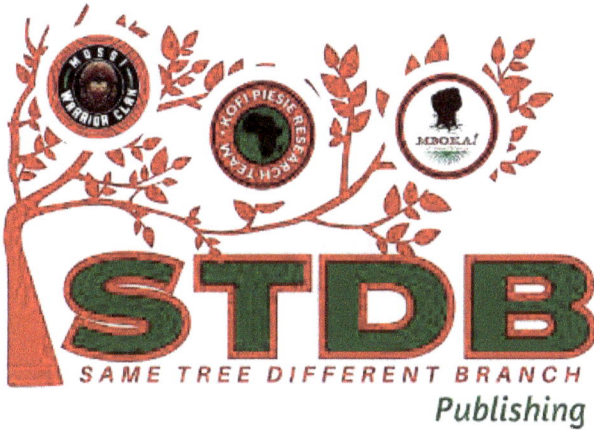

https://www.sametreedifferentbranchpublishing.com/

www.ingramcontent.com/pod-product-compliance
Lightning Source LLC
Chambersburg PA
CBHW071447200326
41519CB00019B/5648